The Real Heal

Genesis Code Vs. Genetic Code: Why Medicine Can Never Divorce Energy From Healing

I0478631

by
Amanda E. Soulvay Plevell, CNHP
Doctorate of Clinical Naturopathic Medicine
PhD-Doctor of Philosophy in Natural Medicine

Abstract

We find ourselves as never before at a pinnacle of healthcare for all humanity. Not only is it seen daily that current systems and models of thinking beg for more truth, accuracy, and result, but because of this lack are more and more people looking to alternative and/or complementary options.

More and more people are also reporting more and more illness and less and less happiness. Drug use, both street and pharmaceutical, is an epidemic. Depression, anxiety, obesity, and nutritional diseases are also epidemic.

Science still wishes to separate and exclude even though people are taking it upon themselves to take their health matters into their own hands. If we want safe, effective regulations of health care, we must look again to the age old conflict between science and creation, and how it has been proven over time, theory after theory, that science and theology, or energy, cannot be separate.

This paper summarizes some of the many theories that have pointed arrows to the eventual result: that Science cannot divorce energy from healing. The health of the world and its inhabitants depends on it.

The tenets of the premise:

1. True wellness combines a stable energy, which would be incapable of producing resultant vibrations and symptoms of "un-wellness", combined with a perfect flow of that energy stemming from a functional "blueprint" that is mapped inside each cell.

2. There is a blueprint in the cells made up of all thoughts, emotions, experiences, and everything that has ever been.

3. This blueprint is composed of matter and energetic factors including thought, emotion, and creation itself.

4. There is no way that individual cells, and thus human bodies with their resultant health expressions could create something anew, apart from the electrical, energetic, and particled constructs of the entire universe, from the electrical source of creation to present day. There is indeed not only an innate system

of organization in the human body, but that there is a function and purpose to all things that exist.

5. That the makeup and memory of those cells INDEFINITELY has to have an effect on the biomechanics of the physical body,

6. and that there can only be true healing with these tenets in mind.

7. Existentialism is necessary to healing.

8. There is indeed not only an innate system of organization in the human body, but that there is a function and purpose to all things that exist.All things have a purpose and a function in accordance with the good of the whole, including situations, events, thoughts, ideas, and emotions. All are creation.

9. A purely physical approach to an individual's apparent state of "health" is a gross error in representing it's true state of "wellness".

10. Emotional energy and disease energy has a viable and necessary place in cellular health.

11. If healing is explained and accomplished purely in a physical construct without a metaphysical energy, then the birth and death of the cells should be able to explain and prove healing.

12. We are influenced not only as individuals, but by a collective and continuous energy.

13. Not only WHAT the mind is thinking but the very ENERGY that is created correlating to those specific thoughts is just as important in our study of healing.

14. Not only is our cellular patterning based in Genesis Coding(creative conception; individually and universally), but also in Conceptual beliefs, thoughts, and ideas.

15. Becoming aware of the activities of the mind can be powerful in altering the code the cells operate off of.

Table of Contents

Introduction

The world has been forever studying that which can or cannot be provable, the very definition of the word "science". Accepted application of the proven scientific principles is the practice of medicine. According to Medilexicon's Medical Dictionary, "According to Medilexicon's Medical Dictionary, Medicine is 1. A drug. 2. The art of preventing or curing disease; the science concerned with disease in all its relations. And 3. The study and treatment of general diseases or those affecting the internal parts of the body, especially those not usually requiring surgical intervention." [1]

The unfair and untruthful suggestion in this definition is that definition of the health of the body is a result only of the condition of the physical body itself. This is the

---

[1] Team, The MNT Editorial. "What Is Medicine? A History Of Medicine." *Medical News Today*, MediLexicon International, 5 Jan. 2016, www.medicalnewstoday.com/info/medicine.

conventionally accepted idea in the practice of health and wellness. While this theory definitely holds water, particularly in the treatment of acute conditions, is it the end all when it comes to health, being and feeling well, and happiness? "Is there more to it" has been a question that has been raised time and time again throughout the history of science and medicine.

If "drugs, prevention and cures and the study and treatment of general diseases" were the answer to complete and total health, then it would work for every one, every time. And what of the so called "miracle" cases of healing, those that defy scientific reasoning? The question arises, then, what else is behind illness and disease, if Louis Pasteur's Germ Theory is not the only precedent? We continue the study that theologians, scholars, researchers and philosophers have attempted to answer this very question at numerous places in history. Can medicine be separate from spirit, from energy, from a source?

It is the premise of this article that it cannot. Not only that the two cannot be kept separate, but we attempt to prove that it not only takes a physical and direct acute correction of injury or illness to "heal" disease, but that true wellness combines a stable energy, which would be incapable of producing resultant vibrations and symptoms of "un-wellness", combined with a perfect flow of that energy stemming from a functional "blueprint" that is mapped inside each cell. This blueprint is composed of matter and energetic factors including thought, emotion, and creation itself.

Understanding the laws of physics, there is no way that individual cells, and thus human bodies with their resultant health expressions could create something anew, apart from the electrical, energetic, and particled constructs of the entire universe, from the electrical source of creation to present day. That the makeup and memory of those cells INDEFINITELY has to have an effect on the biomechanics of the physical body, and that there can only be true healing with this in mind. The author has dubbed these the "Conceptual Cell", which working together complete a "Genesis Code" in contrast to a "Genetic Code" . With an understanding of these is the body finally able to heal in truth, entirely and permanently.

Contributing Theories

The Law of Conservation of Energy

To put it simply: everything affects everything. And nothing affects everything. From the time of creation, the energy that created all put into motion the particles that will create everything forevermore. "The Law of Conservation of Energy states that energy cannot be created or destroyed. In other words, the total energy of a system remains constant."[2] Human beings, and bodies, are not separate from this Law of Energy and were created with the same energy that has been transmuted over time, time and time again. In addition to this energetic science, is the second part of the premise: that there is a blueprint in the cells made up of all thoughts, emotions, experiences, and everything that has ever been. This means that the more concepts one develops around any idea, the

---

[2] C., German. "Kinetic Energy and Potential Energy." *Kinetic and Potential Energy | Wyzant Resources*, www.wyzant.com/resources/lessons/science/physics/kinetic-and-potential-energy.

more separate one will be.   The more one has separated himself by creating ideas and attachments to the ideas, the further away from the source of energy he will become, and the more symptoms he will experience. This includes not only those thoughts, energies, and experiences he consciously knows, but those the subconscious has held as an individual, AS WELL AS those collective energies that have not been transmuted.

It would be wise at this point to lay out a pathway of the theories that have thus far contributed to the theory of the Genesis Code and that of the Conceptual Cell.

Micro and Macro Evolution

Evolution encompasses changes of vastly different scales and an understanding of the principles of micro and macro evolutions to our purpose would be necessary at this point. "Microevolution happens on a small scale (within a single population), while macroevolution happens on a scale that transcends the boundaries of a single species."[3]

We, as individual human bodies are a microcosm of evolution, to the world and universe which is macroevolution. The understanding applies to a whole species set as it does to the comparison of individuals to the universe. Everything that affects the universe affects the individual. Likewise, everything that affects the individual affects the universe. This is the significance of the "ripple effect". Not a single action can be undertaken without causing great scales of affectation as it ripples down.

Understanding this, we can also understand that our human physical bodies are no different from any other natural resource and its efforts in sustenance, stability, balance, and natural order, in that even the body itself is used to filter both

---

[3] Understanding Evolution Team. "Evolution at Different Scales: Micro to Macro." *Evolution at different scales: micro to macro*, Understanding Evolution, evolution.berkeley.edu/evolibrary/article/evoscales_01.

in a physical way and also energetically over our mental, spiritual and emotional constructs. Essentially, as we heal individually, we change the world. As we degenerate, so does all of the world.

Mitochondrial DNA

Medical Science as well seems to support this statement, at least on the level of physical evolution, when John Neustadt and Steve Pieczenik stated in their book, "A Revolution in Nutritional Biochemistry" that, "Current theory holds that mitochondria are the descendants of bacteria that colonized an ancient cell between one and three billion years ago."[4] That these mitochondria contain their own DNA[5] is cited as evidence for the theory that mitochondria evolved from

---

[4] Spees JL, Olson SD< Whitey MJ, Prockop DJ. Mitochondrial transfer between cells can rescue aerobic respiration. *A Revolution in Nutritional Biochemistry,* PNAS. 2006; 103(5): 1283-1288

[5] "Mitochondrial DNA - Genetics Home Reference." *U.S. National Library of Medicine*, National Institutes of Health, ghr.nlm.nih.gov/mitochondrial-dna.

free-living bacteria. It is universal knowledge that what once was, still has an immediate effect on what is today. All of this is summation to the premise that 1. The cells of the individual body are INDEFINITELY affected by the genesis of all of creation, and 2. That these experiences and energies through both thought and form create a blueprint on which the body functions and operates.

Biomechanics

Biomechanics is an important field to study when we want to understand the physical and mechanical functioning of the human body as well as the conceptual body. Biomechanics refers to the functioning of all life systems, including but not limited to cells, tissues, organs, and systems. Biomechanics can refer to the functioning of any life form. It is important to the study of the Conceptual Cell due to the belief that while biomechanics is mechanical in nature, it attests to the idea that all things have a function, a purpose, an innate-ness, if you will. If this is so, it seems then to indicate the potentiality and the apparent necessity of a divine system of organization. This is one physical science that shows that while purely physical in nature, it focuses only on the physical, even though it hints at an innate system of organization, it has not been quantifiably studied as a possibility within the medical community.

Quantum Biology

Quantum Biology, however seeks to go further and begins to answer the seeming lack of explanation, outside of an energy theory, in chemical processes of the body like electron and hydrogen transfer.[6] With nothing but an energetic "pull" does the transfer happen. It seems that what the history of science and medicine has shown is that what begins as the study of a physical nature, tends to move to the spiritual, psychological, theoretical, or all of the above.

According to Dr. Samuel Hahnemann in his infamous work, "The Lesser Writings of Samuel Hahnemann", "You cannot divorce medicine and theology. Man exists all the way down from his innermost spiritual to his outermost natural."[7]

---

[6] Dutton, P L, and C C Mosser. "Quantum biomechanics of long-Range electron transfer in protein: hydrogen bonds and reorganization energies." *Proceedings of the National Academy of Sciences of the United States of America*, U.S. National Library of Medicine, 25 Oct. 1994, www.ncbi.nlm.nih.gov/pmc/articles/PMC44996/.

[7] Hahnemann, Samuel. "The Lesser Writings of Samuel Hahnemann" , republished, Ann Arbor, MI, University of Michigan Library, 2011.

While Biomechanics studies on the mechanical movement of the physical life form along with its relationship to its environment, Functional Wellness goes a step further by including the functioning purpose of every cell and system. Again, it speaks to a suggestion of an innate and divine sense of organized structure. If all things function with a direct purpose, the perfection of that functioning would seem to lead to a perfectly ordered system.

Functional Wellness

Functional Wellness is understanding the functioning of each individual part and its perfect functioning in relationship to the functioning of the whole. If all parts are playing their role, serving perfectly in their function, the body is capable of performing to its optimal performance. It is able to achieve "wellness".

Whereas, the definition of health that is conventionally accepted is generally accepted as "the absence of disease", "wellness" is the optimal functioning and performance of the body *despite* a presence of disease or symptoms.

Holistic Wellness

Holistic Wellness is the optimal condition of all dimensions that make up a human life and factions in which they participate including their Social, Intellectual, Environmental, Financial, Physical, Spiritual, Occupational, and Emotional interactions. Holistic Wellness finally begins to pull the physical action, the outside environment and its influences on the body together with the metaphysical wellness of the spirit, emotional, and mental bodies.

While we have shown that there is and has been the contrasting views of physical medical biology and its counterpart: the metaphysical being-ness of an innate design, thus far, these theories have largely pointed to a separate and INDIVIDUAL view of health. At most, popular theories have allowed the agreement that each INDIVIDUAL is a PART OF a greater metaphysical source, but NOT an understanding that incorporates existentialism as necessary to healing.

Divine Functioning

As the macro is to the micro, so is there a mirror from one set of life systems to the next. Just as the the whole body relies on the proper functioning of the individual parts of the body, so does the great cosmos depend on the functioning of the individual expressions of its divinity. Specifically, this article's purpose is concerning the human individual. The good of the whole is determined by the proper functioning of the individual fractions of the divine personality playing their parts, performing their functions, a literal mirror to the precedence of Biomechanics and Quantum Biology. For the purposes of this article, we will call this Divine Functioning. This is the recognition that there is indeed not only an innate system of organization in the human body, but that there is a function and purpose to all things that exist. On an individual level, Divine Functioning is the ability of the individual to see, to know, and to perform the soul's purpose as an expression out into the world.

Satori

Identifying and performing this is an act of Satori. Satori is the perfect moment of complete and utter clarity. It is the state of peace the human psyche attempts to achieve to feel bliss, joy, and happiness in their sense of complete knowingness. Satori is that state of Oneness, of being at one with Source energy, with the great cosmos pool in which each individual expression originates from and returns to. In this perfected state of knowing and being in a state of Oneness, the individual feels connected, and from this source connection can move forward with the power that source energy provides in it's all knowingness and the person is at peace in their right mind.

Satori is the state of Atonement, a return to creation, and being one in truth with the Creative Energy. If Satori is at one with creation, it would follow that the body would re-align to a state of perfect functioning, as its place in a world of divine energetic and reliable order.

Satori is the understanding that man, in all forms and conditions, came from truth and wholeness, and, as Dr. Samuel Hahnemann was known to say, "...as long as man continued to think that which was true and held that which was good to the neighbor, that which was upright and justice so long man remained free from disease, because that was the state in which he was created."[8], it can be proven that truly that love is the key. A layman's version of Philippians 4:6-9 which lays out an exact plan of how to live, in a way to connects to the art of Satori, love, and divine functioning.

"6 Fret not about anything, but in everything, by prayer and supplication with thanksgiving, let your requests be

---

[8] Hahnemann, Dr. Samuel, "The Lesser Writings of Samuel Hahnemann".

made known unto God.

7 And the peace of God, which passeth all understanding, shall keep your hearts and minds through Christ Jesus.

8 Finally, brethren, whatsoever things are true, whatsoever things are honest, whatsoever things are just, whatsoever things are pure, whatsoever things are lovely, whatsoever things are of good report, if there be any virtue and if there be any praise, think on these things.

9 Those things which ye have both learned and received and heard and seen in me, do; and the God of peace shall be with you"[9]

But how does that play into our physical health? Simply the fact that all things that take away from goodness,

---

[9] "BibleGateway." *Philippians 4:6-9 KJ21 - - Bible Gateway*, www.biblegateway.com/passage/?search=Philippians%2B4%3A6-9&version=KJ21.

wholeness, truth, and happiness are against the very nature of creation. Any efforts to dabble in these things take away the building of health.

"Only human beings have lost this original condition of the consciousness, and have thus become more complicated. When we start trying to explain what Satori is, we start putting it into limiting categories, saying, 'Satori is this, Satori is not that.' In Zen, there are no categories; nothing can be separate."[10]

In medicine, in study of the structure and actions of the body, too, separating and categorizing and taking it apart bit by bit only seeks to separate further, moving against the very true healing, which is a functioning oneness.

It is the purpose of this article to imply that a purely physical approach to an individual's apparent state of "health" is a gross error in representing it's true state of "wellness".

---

[10] www.zen-buddhism.net. "Satori or Awakening." *ZEN BUDDHISM | Zen Concepts | Satori*, www.zen-buddhism.net/zen-concepts/satori.html.

The biological sciences will trace back all the way through evolution to prove the physical structures and changes along evolution's path, but have yet been unable to pinpoint the unidentifiable and unexplainable.

Creationism

That which is unidentifiable and unexplainable has been what has puzzled and been a source of debate for philosophers, scientists, doctors, and theologians for centuries. Some have developed and accepted the idea of Creationism and plenty of PhD scientists believe that this is not only a valid scientific perspective but that if we use the same design inference as with such disciplines as other scientific studies, "if we applied the same logic to biology, genetics and astrophysics it would be evident that not only all life on Earth but the entire universe itself was created by some sort of Intelligent Agent outside of our realm (i.e. God)."

Creationism has been interpreted on both ends of the coin, and has no set designation whether or not it is scientific or metaphysical. It is not monopolized by religion, and if you follow the basic theory that science is an the impartial investigation of a subject using observable evidence, one might find that there is or is not enough evidence to say creationism is a science, itself, one that explains the beginning of time. What comes out of it is what many have been unable to answer: that there is no definable truth for all beliefs on how exactly life began: out of matter, or out of nothing (God or some energy source). This is further support of our premise that much of what is created is truly our perception of it.

The Law of Attraction

The Law of Attraction states that we attract those vibrational energies that match the vibrations with which we put out. This exhibits a demonstration of magnetic laws of physics and is the very core of earth itself. Physics cannot be discussed without physics properties of magnetism, and neither should any study in correlation with physics, the least of which, the human cell. All things must comply with physics law, including forms of energy in thought. Thoughts therefore must, by law, also comply with the physical Law of Attraction and the vibrational pattern of the energy of a thought, the energy of the person having those thoughts must also be applying this basic law of physics.

Even though law of attraction explains magnetics and magnetic energy, it still does little to explain how those energies began in the first place, unless magnetics did simply just attract magnetic particles while spinning in orbit until the mass was great enough to cause compression and the creation of other things.

In the theory of the Conceptual Cell, it is worthy to understand the origination of all things, for as the micro is to the macro, it can do much for telling us how anything mutated from one energy -turned -form into the next.

Biogenesis

Biogenesis delineates a newer perspective about how life regenerates, "as people as late as 150 years ago believed that life came from inanimate materials. For example, because mice and flies were abundant near straw and meat, people thought they came from straw or meat."[11] "Louis Pasteur proved spontaneous generation was false in the 1860s. Now, scientists see biogenesis as the only natural source of living things"[12]

The definition of biogenesis states that it is, "the development of life from preexisting life." "Biogenesis refers to the production of life from life. The production of life from non-living matter is abiogenesis. Abiogenesis has never happened naturally and there are no accepted models for how it could happen in a laboratory or

---

[11] "What Is Biogenesis?" CompellingTruth.org, www.compellingtruth.org/what-is-biogenesis.html.

[12] "What is the meaning of creation 'ex nihilo'?" CompellingTruth.org, www.compellingtruth.org/creation-ex-nihilo.html.

elsewhere. Biogenesis occurs constantly, from the division of bacteria, to trees shedding seeds, to cows calving."[13]

But it does not say how biogenesis originally started. With all of the medical and scientific theories of healing, none of them give a plausible explanation of how biogenesis started. It seems to make sense that we should have an understanding of original biogenesis if we want to have an understanding of how to heal. How a substance is created can give much knowledge to how that substance can return to a state of perfect origination once again.

The term also fails to describe the process of God forming life out of nothing (Genesis 1:1) which is referred to as "creation ex nihilo", literally meaning "from nothing" in Latin. Biogenesis is also different from "creation ex materia, the formation of God creating Adam out of dust, as spoken in Genesis 2:7.

From a standpoint of the first law of physics, what

[13] "What is biogenesis?" CompellingTruth.org, www.compellingtruth.org/what-is-biogenesis.html.

seems very puzzling is that we know by the first law that energy can neither be created nor destroyed. "The first law of science is that the state of matter can be altered, going from solid to liquid or gas and back again, or that atoms can be combined or split. If God really created everything from nothing, and "all things" includes matter itself, we are left wondering very legitimately how the Bible could possibly be telling us the truth. Creation "ex nihilo" is a supernatural occurrence."[14] "The key to grasping the concept of the supernatural is to allow your mind to accept the idea that everything observable is not actually all there is to know. The set of "things that are possible" is larger than the set of "things that we see or experience as human beings."[15] Energy allows us to use evidence other than our five physical senses which are the only sources for attainable information, whereas Science restricts to only what is observable through the five

---

[14] "What is the meaning of creation 'ex nihilo'?" CompellingTruth.org, www.compellingtruth.org/creation-ex-nihilo.html.
[15] "What is the meaning of creation 'ex nihilo'?" CompellingTruth.org, www.compellingtruth.org/creation-ex-nihilo.html.

main senses. "The supernatural is beyond our powers of observation, and is therefore beyond the power of science to explain"[16]

What seems to be interesting is that researchers and philosophers want to separate the two: science and theology. When in actuality, much of what has been discovered through scientific theory and process works to EXPLAIN spirit. Along with the First Laws of Physics, the Laws of Thermodynamics also support a conclusion that an energy, or spirit, or force outside of the physical exists.

The Laws of Thermodynamics

"The first law of thermodynamics involves the conservation of energy. Most importantly, for the sake of creationism, this law states that energy can neither be created nor destroyed; it can only be changed in form or function. This means that the total energy of an isolated system is constant,

---

[16] "What is the meaning of creation 'ex nihilo'?" CompellingTruth.org, www.compellingtruth.org/creation-ex-nihilo.html.

neither increasing nor decreasing. This is evidence for creationism on the basis of simple logic. If no natural process can create or destroy energy, then neither the universe nor the laws of physics can explain the existence of energy. In other words, if energy is impossible to create, why does it exist? The most reasonable explanation is something, or Someone, outside of the laws of physics, and outside of the universe."

The second law of thermodynamics consists of a concept known as entropy. This principle defines the natural and innate trend of systems to move towards destruction or imbalance. "The second law of thermodynamics provides evidence for creationism in several ways, but is often misunderstood and therefore misapplied. As with the first law, the concept of entropy leads to some logical conclusions about the universe. If the universe was infinitely old, it would now be in a state of maximum entropy: of total chaos. All temperatures in the universe would be equalized, and there would be nothing but a formless fuzzy oblivion. Since this is not the current state of the universe, the universe must have a finite age, and therefore a beginning. This makes it reasonable to consider the universe an "effect," not a "cause," and therefore it requires something beyond it in order to begin"[17]

The First Law of Physics

"If all of this is true, and it seems most people do in fact believe in the laws of physics, then it must also be, at least in following the same logic, that someone or something outside of the laws of physics is responsible for creation and origination."[18] If this is the case, then science alone cannot heal on its own. And further proving the Conceptual Cell Theory, the very first law of physics states that energy can neither be created nor destroyed. This proves that not only does SPIRIT or ENERGY need to be considered in healing,

---

[17] "How do the laws of thermodynamics provide evidence for creationism?" CompellingTruth.org, www.compellingtruth.org/laws-of-thermodynamics.html.

[18] "Does The Soul Exist? | BEYOND BIOCENTRISM / Robert Lanza, M.D." Beyond Biocentrism, 3 Dec. 2017, beyondbiocentrism.com/does-the-soul-exist-evidence-says-yes/.

but that EVERYTHING that has been created since the beginning of time, must in effect be considered as to be part of the energy of each individual cell.

Theistic Evolution

Theistic evolution tries to bring together the best of both beliefs; that there is the existence of a creator, but also that once this power God created his world, he then created the laws that would govern that world.

"There are two basic schools of theistic evolution. The minimalist view differs from atheistic evolution only on one point—the beginning of creation. It teaches that God first determined and established the physical laws of the universe with a mind to eventually develop human life. Then He initiated the Big Bang (or whatever phenomenon marked the first moment of our cosmos). After that point, when it came to the development and care of life on Earth, God stepped back and let His laws dictate what would happen. About half of all

Catholics, Orthodox, and mainline Protestants hold this view, as well as most Buddhists, Hindus, and Jews".[19]

The question now becomes: What ARE the laws that dictate what will happen? How does that affect the idea of health and healing when it comes to the very physical scientific, as well as what now also MUST be considered; the cellular energies within? What appears to be happening in this attempt is the same as when philosophers, researchers, and theologians attempted to identify the original creation: that it is either SCIENCE ALONE, or ENERGY ALONE that is the healing curator. Both must be considered in effort together. But the fact remains that NEITHER ONE identifies that original energy(the Genesis Code/the Conceptual Cell) is an essential part of truth in healing.

---

[19] "How do the laws of thermodynamics provide evidence for creationism?" CompellingTruth.org, www.compellingtruth.org/laws-of-thermodynamics.html.

So what have been the theories of the scientific laws of healing; the laws that will dictate what will happen with the world of God?

Contributions from Science

The Germ Theory of Disease

"From the time of Hippocrates, healers conjectured about the possibility of invisible organisms causing disease."[20]

"The Germ Theory of Disease states that many diseases are caused by microorganisms. These small organisms, too small to see without magnification, invade humans, animals, and other living hosts. Their growth and reproduction within their hosts can cause a disease." [21] If this were entirely true, then everybody everywhere that was exposed to germs would get sick and that simply is not the case. Even the theorists understood that disease couldn't entirely be the result of germs, " 'Germ' may refer to not just a bacterium but to any type of microorganisms, especially one which causes disease, such as protista, fungi, virus, prion, or viroid. Microorganisms that

---

[20] http://www.wholehealthnow.comhomeopathypro/miasms-03.html
[21] https://en.wikipedia.org/wiki/Germ_theory_of_disease

cause disease are called pathogens, and the diseases they cause are called infectious diseases. Even when a pathogen is the principal cause of a disease, environmental and hereditary factors often influence the severity of the disease, and whether a particular host individual becomes infected when exposed to the pathogen."[22]

Another incongruency is the evolutionary study. Where did the "first germ" come from?
Epigenetics allowed some answers to this potential threat to the germ theory.

Epigenetics

"The followers of the spontaneous generation theory believed that germs appeared whenever the conditions were right for their development without the need for reproduction. In some sense this is true as where did the "first germ" come from?"[23]

---

[22] https://en.wikipedia.org/wiki/Germ_theory_of_disease
[23] "Miasms in Classical Homeopathy." What Are Miasms?,

The zymotists suggested that they weren't germs at all, but certain substances called "zymes" that certain patterns or substances could lie without effect until the time when the conditions and environment were ripe, thus causing disease. This is belief of the early 18th century theory of medicine. This sounds like epigenetics, that something is latent and only present when the conditions are ripe for it being so. Why would it be any different from latent patterns from centuries ago? The only difference is that these theories discussed it as being latent in one particular body, which differs from the Genesis Code, which alerts that it is possible this latency that could be awakened, could be so from from years and centuries ago.

Hahnemann's Miasm theory however postulates closely to but not reaching the conceptual cell theory in totality.

The Miasm Theory

In Hahnemannian Homoeopathy, the word "miasm" means "the effects of microorganisms on the vital force including the symptoms that are transmitted to the following generations. These chronic miasms are capable of producing degenerative illnesses, auto-immune diseases and lead the organism toward immuno-deficiency disorders."[24]

"Hippocrates was the first physician to use the term 'miasm', which has its origins in the Greek word for "taint" or "fault."[25] "He postulated that certain infectious diseases were transmitted to humans by air and water tainted by miasms."[26] In this understanding, differing from Hahnemann's theory, is that the "miasm" could be a pathogen and not necessarily an energetic currency through cellular mutation. This corroborates our premise that emotional energy and disease

---

[24] "Where Kent Differs with Hahnemann - Page 4 of 4 - COMPLETE Information about Where Kent Differs with Hahnemann - Page 4 of 4 - David Little." Homeopathy, 28 Dec. 2009, hpathy.com/organon-philosophy/where-kent-differs-with-hahnemann/4/.
[25] "Inherited Tendencies." Center for Homeopathy, www.centerforhomeopathy.com/blog/inherited-tendencies.
[26] "Miasms in Classical Homeopathy." What Are Miasms?, www.wholehealthnow.com/homeopathy_pro/miasms-03.html.

energy has a viable and necessary place in cellular health.

Current theories are still based on a scientific, followable paper trail evidential train. This is expressed in nearly every medical textbook, even in natural health and integrative medicine as is expressed in the statement that. "Nutrient insufficiencies affect multiple organ systems in ways dependent on the history and genetic makeup of each patient…"[27] While true, this statement encompasses popular conventional, logical belief that all disease has a physical answer or marker, focusing only the Genetic Code, when it has already been proven that mitochondria itself has its own DNA, which is ever changing as the mitochondria changes.

No study into the theory of disease could be complete without the life of the cell including its mutations. If healing is explained and accomplished purely in a physical construct

---

[27] Phillips, Theresa. "How Genetic Polymorphism Promotes Diversity and Lasts Over Generations." The Balance, www.thebalance.com/genetic-polymorphism-what-is-it-375594.

without a metaphysical energy, then the birth and death of the cells should be able to explain and prove healing.

The Physical Cell

Cells are not just physical, carrying the code for everything that tissue of the body needs, but they, because they move, create a frequency, and as energy also hold the code for any energy that matches their own frequency...the law of attraction. The only way that changes is to establish a new frequency. In order to exist, cells attract and join and develop, growing into masses with traits that characterize it. Cells have a memory and repeat themselves after the pattern that is stored in the individual cell before it. The physical matter that was there in the parent cell repeats itself in the newly formed cell, along with the part that we didn't see...the energy, the essence, all that was part of the adult cell that we were unaware of. In the case of a human parent, it is all the thoughts, emotions, feelings, beliefs, and concepts that were present. The energy, the vibratory patterns of such things was just as present as that which we can see under a microscope, but the human body does not narrow down to such simplistic, provable matter such as science. It is a mystery the essence of which cannot be studied in a laboratory. The essence of which hundreds of civilizations, having developed hundreds of religions have tried to make sense of for centuries. And what they fail to understand, as we said understanding is the reason, is that to understand means to the death. To have all the answers means formless life once again. Life is a formed hiatus for the spiritual body. A chance to play, to experience, to feel, to move, to have the joy of remembering and the reconnection of love and oneness. It IS more than genetics, and genetic derangement certainly must have reason for it's variated sequence.

Genetic Polymorphism

"There are thousands upon thousands of receptors on each cell in our body. Each receptor is specific to one peptide, or protein. When we have feelings of anger, sadness, guilt, excitement, happiness or nervousness, each separate emotion releases its own flurry of neuropeptides. Those peptides surge through the body and connect with those receptors which change the structure of each cell as a whole. Where this gets interesting is when the cells actually divide. If a cell has been exposed to a certain peptide more than others, the new cell that is produced through its division will have more of the receptor that matches with that specific peptide. Likewise, the cell will also have less receptors for peptides that its mother/sister cell was not exposed to as often"[28]

"Genetic polymorphism is understood to be a variation that occurs in a DNA sequence producing an abnormal status."[29] It is essential to study Genetic Polymorphism in the study of

---

[28] "How Your Thoughts Program Your Cells." High Existence, 7 Feb. 2011, highexistence.com/thoughts-program-cells/.
[29] "Gene Polymorphism." Wikipedia, Wikimedia Foundation, 27 Dec. 2017, en.wikipedia.org/wiki/Gene_polymorphism.

healing because of two questions of interest. One being that it does not take into consideration necessary and innate change in evolution as a whole. It seeks instead "where it went wrong" rather than looking at the changes of the entirety of evolution. It seeks to identify genetic impedences without considering the nurture portion of "nature vs. nurture". Again proving that if that theory were indeed true across the board, the environment of which the cell finds itself at its birth must be considered. Proving the Genesis Code and the Conceptual Cell Theory in which everything present at the birth of the cell matters.

The second area of interest that flags attention when discussing Genetic Polymorphism is that while GP seeks to understand "where it went wrong", Conceptual Cell Theory ALSO seeks the same, though searching instead where a separation occurred through an energetic sequence; a disruption of the connection to the main source universal energy is also examined.

Genetic Polymorphism and Mutations

"Mutation results due to DNA sequence changes specifically that happen once an allele is transferred from one generation to another and initiate alterations in the allele status from normal to abnormal. In contrast, gene polymorphism is defined as a variation that occurs in allele in a DNA sequence"[30]

"Gene polymorphisms are caused by duplications, deletions, and a mutation of triplication of high quantity of DNA base pairs sequences. In addition, Polymorphisms may occur due to changes inside introns or changes in regions for one or multiple DNA bases that are between genes. If the changes occur in a gene' coding sequence, then different phenotypes may appear as a result of protein variation that is caused by sequence changes. These changes are located exactly in genes' coding sequence"[31]

---

[30] Phillips, Theresa. "How Genetic Polymorphism Promotes Diversity and Lasts Over Generations." The Balance, www.thebalance.com/genetic-polymorphism-what-is-it-375594.

[31] Phillips, Theresa. "How Genetic Polymorphism Promotes Diversity and Lasts Over Generations." The Balance, www.thebalance.com/genetic-polymorphism-what-is-it-375594.

However, a representation again of science proving against itself as acting alone in favor of the study of metaphysics as part and parcel is the fact that it has been identified that, "Only about 5 percent of gene mutations are thought to be the direct cause of health issues. That leaves 95 percent of genes linked to disorders acting as an influencers, which can be influenced one way or another, depending on life factors. Of course, many of these are beyond one's control, like childhood events, but some are entirely within one's control, such as diet, exercise, stress management, and emotional states. The last two factors are directly dependent on one's thoughts."[32]

---

[32] Hampton, Debbie. "How Your Thoughts Change Your Brain, Cells and Genes." The Huffington Post, TheHuffingtonPost.com, 23 Mar. 2016.

Metaphysical Contributions

The Non-Material Sciences

Science is not the only entity to have developed coding theories. There is much research and study to prove energetic imprints in multiple metaphysical theories as well. Both sides believe that whether they be energetic imprints, or genetic coding, what IS agreed upon between the two separate and opposing components of Science Vs. Metaphysics is that the body is bioelectrical in natural. Not only that the body is a bioelectrical vessel, but that "on a physical level is all the electrical signals running back and forth, most of which is happening below your conscious awareness."[33] Much happens below the level of conscious awareness, and if this is true, then subconscious concepts, thoughts and beliefs need to be

[33]Sassandahalf, et al. "How Your Thoughts Change Your Brain, Cells, And Genes." The Best Brain Possible, 1 Dec. 2017, www.thebestbrainpossible.com/how-your-thoughts-change-your-brain-cells-and-genes/.

examined in an effort to understand the body as a whole in order to learn what is directing the electrical signals in the first place. "In the context of observing the micro to the macro, it is interesting to note that just as the debate continues of creation vs science, we can follow it with a debate between the physical body vs. the energetic body as being the regulator of what was, from either theory, the original construct."[34]

It would behoove us to remember Constantine Hering's Law of Cure, which demonstrates an understanding of how the the macro cares for the micro, meaning all that happens on the large scale affects the small, and vice versa. "Hering's Law of Cure is the basis of all healing...This is the way the body heals or cures itself. All cure starts from within- out, from the head-down and in reverse order as the symptoms have appeared or been suppressed."[35] This latter shows us that the Genesis Code

---

[34] maudstonge.com. Articles on Homeopathy from Master Homeopathic Practitioners, www.homeopathy.ca/articles_det17.shtml.
[35] "Hering's Law of Cure." Healing Naturally by Bee, www.healingnaturallybybee.com/herings-law-of-cure/.

understands the progression of all events that have occurred are going to have a direct affect on the body's decision to heal.

Alongside the study of biological sciences, and in an attempt to gain understanding on the basis of true, total, and whole wellness are the non-material sciences, under which Quantum Psychology falls.

Quantum Psychology

Quantum Psychology offers a practical approach to becoming aware of our automatic responses and how they influence our behaviors. "Quantum" is the term used for the metaphysical energy that is studied. "Psychology" is the study of the mind and thought resulting in behavior. Put the two together, and you get the study of the energy of the mind and thought processes that direct, consciously and/or subconsciously, the behavior and actions of the functioning being. While apparently sounding of the New Age Genre, and not accepted as a study in the medical sciences, it has been thought and therefore must be looked at if a rationale for total wellness is to be discovered. The study of such becomes an either/or scenario, rather than a glimpse at how perhaps both methodologies contain within them some truth. Especially because the question resorts back to the original debate of "Nature vs. Nurture", posed by John Locke. Obviously, both the former and the latter create an effect on the human being and psyche. While it lends only partially to the premise in that there is more to health than medicine, it did not come close to indicating that we are influenced not only as individuals, but by a collective and continuous energy. Also to be noted is that this contemplation dates as far back as 1690[36], much earlier than the practice of modern medicine, which began in the 1800's[37]. It also is interesting to note that John Locke's view of a "tabula rasa" or "blank slate" in which we are a blank

---

[36] https://www.coursehero.com/file/19960291/nature-vs-nurture/
[37] Team, The MNT Editorial. "What Is Modern Medicine?" *Medical News Today*, MediLexicon International, 5 Jan. 2016, www.medicalnewstoday.com/info/medicine/modern-medicine.php.

canvas onto which our experiences are painted and otherwise did not exist before, was in direct criticism of Rene Descartes' view of a universal God that existed to all humanity. It seems the debate that most notably first started in the 1600's would not be answered, but continue to be a cause for question even into our present day. Today, as in it's origination, the question seems to ask how heavily is functional wellness, the functioning of the body systems, the biomechanics, or the function of each cellular and systemic unit in the makeup of a life influenced by the concepts, the ideas, the beliefs in which it subscribes.

Quantum Theory

Quantum Theory is the idea, stated by Planck, that "Physical systems can only possess certain properties, such as energy and angular momentum, in discrete amounts or (quanta). Stemming off of this original idea are DeBroglie's wave mechanics theory and Heisenberg's Uncertainty

Principle." [38] All of which contribute to the belief that even

the most minute quanta of energy has a direct ripple and

graduating affect on future generations of energy. Therefore,

the study of even the tiniest fraction of energy is bound to

have an impact on everything else, time immemorial.

---

[38] Quanta and Wave-Particle Duality - Quantum Theory and the
Uncertainty Principle - The Physics of the Universe,
www.physicsoftheuniverse.com/topics_quantum_quanta.html.

Quantum Mechanics

Quantum Mechanics is the study of how this theory is used. Many researchers and believers in Quantum Theory and Quantum Mechanics have toyed with the modes of how healing may be possible using energy and various forms of energy healing. QET, for example, claims to shift energy by a restructuring at the cellular level, ultimately "engaging the body's innate ability to heal"[39], this not being the first to mention this concept. QET goes on to express that "the foundation of QET is built with the understanding of Quantum Physics. The body stores information as packets of energy in the cells, almost like a cellular journal. It's like our brain notes down everything we experience into our cells. Some of these memories are useful while other memories limit our lives or cause 'dis-ease'. When processed, all of the

[39] "What is Quantum Energy Transformation™ (QET™)?" *What is Quantum Energy Transformation™ (QET™)? |*, www.quantumhealingcenter.com/quantum-energy-transformation.

information (energy) becomes useful. We get to keep learning and lose the limitations."[40]

Quantum Biomechanics

Less scientifically quantifiable is the study of quantum energy and yet it is the stuff the very world is made up of, including ourselves, according to the theory of Quantum Biomechanics. Quantum Biomechanics refers to the energy of the movement, function, and purpose of all life. This portion of the theory pertains to how the study of quantum energy is applied and understood concerning human life. This theory offers up an understanding of the energy it has on the functional wellness of the individual, and its possible changes to that life's biomechanics, experience of symptoms, and thus, health. As we can see by this view, health is NOT just the absence of disease, but there is clearly more present.

[40] "What is Quantum Energy Transformation™ (QET™)?" *What is Quantum Energy Transformation™ (QET™)? |*, www.quantumhealingcenter.com/quantum-energy-transformation.

Quantum Pathology and Psychology

From this, we begin to understand the progression into the world of Quantum Pathology, which serves to understand the consequences and disease states, or distortion of the functioning of the body due to causal relationships to vibrations of energy.

Specifically correlating first to Quantum Psychology, we are looking at vibrations of energy that are potentially or indefinitely created from thoughts, concepts, and belief systems, and how these vibrations of energy manifest in the human body. This, again, is the basics of the Law of Attraction coming into play, as we mentioned previously.

Simply put, Quantum Psychology is the study of the energy of thoughts in the human mind. Essential to our premise is understanding this means not only WHAT the mind is thinking but the very ENERGY that is created correlating to those specific thoughts is just as important in our study of healing. The study of this energy is then useful in Quantum Pathology to study the effects, correlations, and connections which are less evidentially proven through scientific study.

Practitioners of Quantum Energy Healing like QET are not the first to understand and try to bridge the gap of physical medicine to that of the psychic world. Additionally progressive theories take the Quantum Psychology theory and expand into basic rationale and thought processes that experiences ingrain within a person's psyche, and subconsciously portray themselves on the stage of one's life. These vibrational patterns are bound to have an influence on the resulting pathology, simply because of the energy developed. This is the "electrical cord" the energy follows to manifest the "power", otherwise known as the expression of the physical body. There is a uniqueness here, in that it involves energy in all forms, whether or not they are consciously recognized.

Concept Pathology

Concept Pathology is similar to Quantum Pathology in that it involves the theory of vibrational energy coming directly from thoughts and mind activity and its influence on the creation of the physical body. Whereas "Quantum Psychology" is the STUDY of the vibrational energy of the thoughts of the human mind, and "Quantum Pathology" is the vibrational energy itself and its possible causal relationship to manifestation in a physical form, namely, in the human body.

"Concept Pathology" is the THOUGHT ITSELF and it's possible causal relationship to manifestation of physical symptoms in the physical form. Concept Pathology is the spin off term the author uses to take the work of Dr. Thurman Fleet a step further through working to study the possibility of cellular healing requiring an understanding of Genesis itself.

Men like Dr Thurman Fleet have made substantial contributions to the study of Quantum Theory, Energy, and Biomechanics. As a researcher of all things energy, he formed a theory of conceptual processing in healing called Conceptology or Concept Therapy.

Concept Therapy

Dr. Thurman Fleet was a man of vision. Every day more and more people search for truth, for a real understanding of life, and for a natural way to improve their health and their own lives. Understanding this, and also understanding that "there could be no healing without teaching", Dr. Fleet formed his Conceptology and Concept Therapy Programs. Believing in the natural laws that exist for everyone, he set out to educate anyone who was willing to listen about these laws and how to use them to better themselves. In the 1930's he began to introduce and educate chiropractors to these facts. He went on to write and publish his book, "The Rays of the Dawn" in an effort to teach people how to overcome the negative energy forces and transmute them into positive healthy energy. Eventually, he developed seven phases of Conceptology that the seeking student can participate in, and opened the Concept Therapy Institute in San Antonio, Texas to further his mission.

Dr. Thurman Fleet understood this in 1931 when he had "an illuminating experience that resulted in bringing to the world a phenomenal chiropractic healing technique and an unparalleled correlation of the world's Truth, wisdom, and knowledge."[41]

"Concept Therapy is a unifying correlation of the fundamental disciplines (Science, Metaphysics, Theology, Psychology, Sociology, and Philosophy)."[42]

According to Dr. Fleet, "Here, concept means fixed ideas in the subconscious mind, and therapy means having healing qualities. Ultimately, Fleet's teachings focus on the healing of the negative and limiting concepts in the subconscious mind. Happiness, health, success, and peace can be the result."[43]

---

[41] "Intro." *Dr. Thurman Fleet*, www.drthurmanfleet.com/.
[42] "About Us." *CONCEPT-THERAPY*, concept-therapy.org/portfolio-item/about-us/.
[43] "Intro." *Dr. Thurman Fleet*, www.drthurmanfleet.com/.

However, again, as is true in the comparison between the either/or perspective of medical vs. metaphysical, those that accept metaphysical theory are still proving an INDIVIDUAL perspective. At best, it proves an existence of the INDIVIDUAL being made up of a same COSMIC SOURCE, but still does not involve an existential, non-biological look at the construction and evolution of the cell itself as a necessity in healing. In other words, it does not take into account a blueprint of the cell that could date back to creation itself.

The Genesis Code

The Conceptual Cell

The Genesis Code begins with our mutual understanding and agreement of the bioelectrical nature of the body. How it receives the impulses is the topic in question.

Quite simply, the body's electrical natural can be compared to a computer system. The computer, the physical matter of the computer, is the shell, the vessel. Without any code or electricity, it has not use, no purpose, no function. However, if one were to send in a programmer, the circuitry stores all of the "code" that the programmer enters. If it wants a space to occur every time one should hit the "space bar", it's going to direct the computer to do so by typing a code. Now the computer performs this function without fault, every time and always, without exception. For every function the computer is to perform, the circuit boards are programmed with the appropriate response the programmer wants to elicit.

Even yet, after all of this programming, nothing happens and the computer is without function if it is not attached to a form of energy that runs through the unit, giving it power to function; namely: electricity.

In broad, crude terms, the human body and its functioning can be seen as such. This is not to downplay the amazing essence and spirit that is divine guidance, providence and order, but in order that fear and misconception can be removed simply by simplifying the processes. We know that of course the life of a human is far more than it's vessel and function. This modality takes that into account and works with the body on every level, physical(corporal), mental, emotional, and spiritual, and other.

The human body can be seen as the form that was built to carry out the functions of the body. These functions are like that of the "delete" action when one presses the delete button on a computer. "This coding for every function in the body is stored in every cell of the body, which makes up every tissue, fiber and fluid in the body."[44] "While science already agrees with this through the genetic code of DNA, and this being physical in nature, The Conceptual Cell Theory believes "coding" is taken in through through the means of and also affects every other level of the body: mental, emotional, spiritual, and other."[45]

"The Conceptual Cell Theory implies the theory of Creational Coding, in which this coding of every cell starts even before conception, at the very idea within the parents' minds of the possibility, the thought, and even the intention of creating a baby. Truly, it really can start back to the beginning of all life with the first single spark, and follows humanity through all creation and evolutions.

The coding of all things are stored in our cells directing their every action performed outside of and without the help of

---

[44] "Wellness." Plevell, Amanda, Natural Health and Wellness for Natural Living, thenaturalsourcecompany.com/wellness.html.
[45] "Wellness." Plevell, Amanda, Natural Health and Wellness for Natural Living, thenaturalsourcecompany.com/wellness.html.

the conscious mind. Every thought, emotion, feeling, idea, and concept and belief is held in the code of the cells. Every action, behavior, incident, situation, and everything that was learned through them is stored in this code. The DNA and genetic makeup is stored. Everything that was within mom and dad at conception is stored. Every energy that went into this creation is stored. Then as the new being ages and evolves through life, everything that happens on a moment to moment, daily basis is stored as new directions. However, the new directions, the new codes don't "replace" the old code, unless effort is made for the old code to be absolved, to be deleted, to be negated. Since this was all a subconscious act to begin with, the conscious mind cannot begin to fathom that this is necessary, nor how to go about it."[46] Instead, the conscious mind gets involved with trying to "solve" whatever reason came up for the old code to be absolved, and instead of breaking the energy of the code (cutting off electricity to it) the

[46] "Wellness." Plevell, Amanda, Natural Health and Wellness for Natural Living, thenaturalsourcecompany.com/wellness.html.

very act of "thinking" on it creates new synapses, new nerve patterns, new cellular codes and the old code still is not broken, in fact with all of the new pathways these synapses created, there is more "highway" for the codes to travel.

"This is where an individual will experience not only a non-healing or non-restoration of balance, but a compounding of seemingly un-solvable symptoms."[47]

Thus, this code is the most important understanding of why the body is performing and functioning the way it is. However, because it is not performed at the conscious level, it is equally hard to recognize what old codes need to be broken.

Second, the energy provided for this code to be performed is equally important. Code cannot be carried out without energy. This energy comes from the thought you give it (good and bad and through your conceptions and beliefs), the energy you put towards it (through action plans, intentions, and attempts at attainment), and the bio-electrical circuitry of

[47] "Wellness." Plevell, Amanda, Natural Health and Wellness for Natural Living, thenaturalsourcecompany.com/wellness.html.

the physical being.

But what, in essence, does the Code contain?

The Code Itself

The Genesis Code is comprised of three things: the Code of Imprinted Human Experiences, The Code of the Physical Cells, and the Essential Cosmic Energy. All three of these necessarily and by law mix together as ingredients in a formula that ultimately provide the blueprint upon which the cellular "factory" of a physical body will produce it's variations. Each cell is produced exactly as the blueprint dictates. These blueprinted cells cannot simply be labeled by the scientific term, "cell", which embodies a physical construct of atoms, protons, neutrons, and electrons, but must, by the Laws of Physics, include the energetic imprints and vibrational constructs that are also part of that biological micro-world.

This is why there is no direct one path or protocol for a specific disease, why people will react differently with one treatment to the next. What science has been chasing to explain and come up with the perfect protocol to "cure cancer", "cure ALS", or any chronic disease that currently doesn't have scientific understanding, evidence, and predictability behind it will not find it without a complete understanding of the Conceptual Cell. The Genesis, i.e. creation, down to the smallest particle, both physical and quant, builds the construct of the Conceptual Cell patterns forming the Genesis Code for the individual.   In this way not only Genetics, but Genesis, supply the main ingredients for the human formula.

The Code of Imprinted Human Experiences

The code is made up of the experiences one has had so far as a human on this earth. Every thought, word, belief, action, event on the part of the self or witnessed by oneself has imprinted itself into the conscious and subconscious mind as an idea, a belief, or a concept. Everything has been stored, including every emotion, every decision, and every decision's alternative. Every energy that has ever influenced. Every energy that one has ever engaged in, been part of, or been around. All of this has imprinted into one's mind and cells and become part of the makeup of the self. Not all of these experiences are consciously recognized, in fact much of the human experiences are subconsciously stored, creating concepts the individual may not even know exist, as Dr. Fleet realized in the development of his theory of Concept Therapy as mentioned previously.

The Principle of Oneness in Relation to the Individual

The idea of oneness, and a one energy source are not new, however, the facts of looking at this principle as an explanation and in an attempt to UNDERSTAND the fact that , if this were indeed, true, there is no way that individual cells, and thus human bodies with their resultant health expressions could create something anew, apart from the electrical, energetic, and particled constructs of the entire universe, from the electrical source of creation to present day.

As evolution progressed, cells started attracting to others, splitting, merging, and developing greater and greater forms with more characteristics. Evolution moved up through the species as good attracted good and better attracted better and more and more characteristics adapted to the needs of each particular species. The older, lower vibratory patterns and thus less evolved (looking back from our point of view today) actions and being-ness of the caveman have been healed and transformed as evolution morphed the vibratory patterns higher and higher attracting greater and greater characteristics and behaviors.

This is not new theory. The I Ching is based entirely on this truth that one is not alone and that an individual's experience is not singular. There is a pattern, a cycle, a rotation of vibrations that exist that recycle themselves time and again. That is why it is no surprise to be able to know where one has been and where one is going based on where he or she is now. Each person's experience is a different flower grown from the same roots. This truth could not exist if you were the only one anywhere that has EVER experienced what you are currently experiencing.

The trick is not to replay the old patterns. But the mind is only ABLE to review old patterns and therein lies the problem. If we see ourselves and our POWER as a mind, we will only ever be able to create that which has already existed. The mind CANNOT elevate us out of strife, struggle, and turmoil. Only a change in frequency can, only an allowance of attraction of all things at this new frequency can begin to make change in what we see around us. If new, essential vibrations, meaning coming from within the knowing, are trusted, these vibrations then are the ones that are continuously being activated and played upon, the spiral begins to spin upwards attracting and creating only vibrations emanating from the same higher elevation. Every experience one will have in their human life records themselves in the mind as concepts, thoughts, ideas and beliefs. These concepts imprint themselves within the cells as Truth because one chooses to take them in, and believing them, have added them to the truth of what makes one up.

The Code of the Physical Cell

Everything that one has chosen to take into the body, whether it be a thought or belief, or a food, drink or substance is coded to continue to make up the body that needs, desires, or craves those substances. It, along with the genetics one gained from parents and family is what has determined the cells and thus, what has determined the outward expression of the body. These are all of the things we are consciously aware of, and have decision making powers over.

Epigenetics

In direct support of what we have concluded so far is the proving of epigenetics. "The fast growing field of epigenetics is showing that who you are is the product of the things that happen to you in your life, which change the way your genes operate. Genes are actually switched on or off depending on your life experiences, and your genes and lifestyle form a feedback loop. Your life doesn't alter the genes you were born with. What changes is your genetic activity, meaning the hundreds of proteins, enzymes, and other chemicals that regulate your cells."[48]

Bits and pieces of epigenetics are starting to be heard, about how the genes of even great-grandparents can affect their descendents. But it all seems complicated after the simple straightforward theories like the germ theory. As we

---

[48] "What Is Epigenetics?" Edited by Steven Dowshen, KidsHealth, The Nemours Foundation, Jan. 2014, kidshealth.org/en/parents/about-epigenetics.html.

have learned that is not the cure all, more people are willing to look into what the direct and often hidden causes are. Epigenetics may have some of those answers.

"Epigenetics is short for Transgenerational Epigenetic Inheritance, and admits the idea that environmental factors such as living arrangements, diet, lifestyle and behavior not only has a direct affect on the people that are exposed to them, as science is now willing to admit , but that it also can affect the health of their descendents, and in ways beyond DNA and genetic pattern and code."[49]

"Epigenetic experts believe that the environmental conditions and life experiences of parents, grandparents, and even great-grandparents can, in a way, flip "on/off switches" on the genes in their eggs and sperm, or the genes of developing fetuses in pregnant women, thus changing the genetic code of their offspring and descendants. In this way,

---

[49] "What Is Epigenetics?" Edited by Steven Dowshen, KidsHealth, The Nemours Foundation, Jan. 2014, kidshealth.org/en/parents/about-epigenetics.html.

new genetic traits can appear in a single generation, and be passed on to kids, grandkids, and beyond."[50]

This is suggestive that stress, for example, can affect genes, causing those genes that affect obesity to be "switched on" and those messages for inherent health can become switched off. This means that what one inflicts upon themselves is now essentially "re-coding" their cellular patterns, which will ultimately predispose their offspring to cellular patterns that otherwise would have encouraged a pattern of health.

Remember the theory of nature vs nurture, where it was debated whether the genes we inherited or the environment, like eating and exercise habits, in which one is raised contributes the most to the resulting disease presence?

In the past our understanding of disease focused on these interactions. If we applied to the traditional theory of

---

[50] "What Is Epigenetics?" Edited by Steven Dowshen, KidsHealth, The Nemours Foundation, Jan. 2014, kidshealth.org/en/parents/about-epigenetics.html.

genetics, it should follow suit just like a map. There should not be so much unexplainable change, as was the case. It should be predictable, and observable in physical manifestations. Epigenetics may be the reason, which involves an alteration in the patterning due to genetic conditions that switch on and off certain genes and make new conditions happen more rapidly.

Without paying attention to all of the cell, ie. the Conceptual Cell, genetics is like a game of roulette, with a mismatch of unexplained genetic occurrences. "Once again, here is proof that we must look farther than cellular biology to include conceptual mapping and genetic organization if we are to understand disease process and true healing."[51]

Understanding this, it is imperative to know that Epigenetics is not just a "new science" left to the researchers, scientists, and doctors. This is an example of how the individual can have a drastic effect on the coming generations

---

[51] "Robert Lanza." Biocentrism Explored, biocentrismexplored.com/

and the world as we know it. With each every day lifestyle choice and action, one is improving or degrading the world and the health of mankind.

It is not known, at least scientifically, how or if one can switch on those genetic switches that have already been switched off or vice versa, but it is obvious that lifestyle habits will have an impact on future generations just out of habit and is a worthwhile effort. Those that learn healthy lifestyle habits can effectively plan for health in the next generation.

Apparently, the effects of epigenetics are not lost on scientists currently researching the possibility. It's exciting to know that, alongside the belief systems of a Conceptual Cell Theory, genetic roots of a disease must be looked at.

"Scientists also are using the concept of epigenetics to develop new approaches to diseases with genetic roots. Drugs are now being produced that act by switching on or off faulty genes in epigenetic fashion. This type of genetic "quick fix" could be a very valuable strategy in the fight to prevent, treat, or cure a number of conditions, such as cancer, diabetes, and Alzheimer's disease."[52]

Many others theorists and philosophers have attested to a study of energy vs science in the nature of healing. The history writers ascribe to Hahnemann the saying that the nature of the child is dependent on the soul condition of the mother at the time of conception.

CG Carus 1779-1869, as well, a german physiologist, sought to understand the structure of the human organism as an outcome of subconscious psychological function.

Epigenetics will not be the last to attempt to prove the improvable.

---

[52] "What Is Epigenetics?" Edited by Steven Dowshen, KidsHealth, The Nemours Foundation, Jan. 2014, kidshealth.org/en/parents/about-epigenetics.html.

The Essential Cosmic Code

The makeup of a being, as we've shown does not simply reside in the individual themselves, but as epigenetic proves, and the theories of Dr. Hahnemann, that there is at least a genetic following that involves more than physical genes, but the choices one makes.

This is not new information, as it goes even as far back to the times of Constantine Hering, noted in his Law of Cure: ""We don't catch diseases, we create them by breaking down the natural defenses according to the way we eat, drink, think and live".[53]

It seems years have gone by where knowledge could have been gathered if this topic was further explored, just as now we are at the crossroads of understanding that we can continue on a path endeavoring to prove scientifically the

---

[53] "Hering's Law of Cure." Healing Naturally by Bee, www.healingnaturallybybee.com/herings-law-of-cure/.

aftermath of cellular instruction or we can travel the path less travelled and give due credence to the possibilities the the Conceptual Cell Theory appropriates.

Starting at the position of the Law of Cure and the new research epigenetics has allowed, we must go beyond. In addition, it must be understood that just as we said before, and as the first law of physics states, we know and can agree on the fact that energy has been pooled and recycled since time as we knew it began. This energy has known and been many things. It is part of what makes the self up. Vibrations are patterns that are recorded, transcribed into history. Not history from the past as we know it, but history on the vibratory patterns of all experiences of life, altogether, intermingled as one, and yet separate. The resonances of these vibrations, just as the waves of awareness on earth, ebb and flow, just like every other cycle, season, and pattern. It's all about timing, it's all about pattern. This is why numbers have special meaning and code. This is why symbols are sought after and killed over. All are looking for the secret...to riches, to uniqueness, to separateness, and power, really. These patterns run themselves through like a mystery, getting to play themselves out again and again, each time, each rotation bringing with it a new uplift, a new higher level of insight.

This inner knowing is the energy of you, the connection back to the original energy source, the essence of the spirit you are composed of, the construct of vibrational patterns that builds the form that one knows as themselves.

The makeup of the individual Genesis Code is the genesis of the human existence: every thing, everywhere, from every time since the beginning of time. It is the rhythm and energy force that brought about creation. It is the coming into being of every energy. It is the origin of all things, the development of species and civilizations. It is the relationships between God and man, between man and nature, and man and man.

It is about creation: everything that has ever been

created and everything that one is determining now to create. What one thinks, believes, and acts on along with the makeup of their physical constructs is the genesis to your each new day. This includes everything that has ever been, thought about, or acted upon by anyone, anywhere, ever. Everything that has ever been, everything that IS now, and everything that will ever be forms the recipe from which existence is determined and thus creates the human experience. It is the Genesis Code.

The Essential Cosmic Energy

The Essential Cosmic Energy is the original code. The creation of coded vibrational patterns inherent in the universe that was created from all life and all life forms. It is the energetic material that all are composed of at the time of their creation. This is the essential energy. The part of one that is truth and being without form, without decision, just being-ness, complete and whole. It is the code of creation from which all things are formed. It is the essence in each of us.

How it affects what we see is the outward manifestation of symptoms. In order for this essence to have movement, flow and experiences, it must take up residence in the vehicle of form. This form is the movement, the muscle, the bone that one sees in the mirror as self. It is what gives one the ability for thought, for feeling, for emotion, for experience. It is the vehicle that transports one from experience to experience. It is the megaphone in which one communicates. It is the representative of one's vibrational patterns in appearance, action, and behavior. The essence is evident in the portrayal of itself through the body, but it is nothing more than energy in form, vibrating itself through matter.

Human beings are a mass of cells floating in water. Every bone, tissue and fiber of our bodies are cells. Some cells are made to be muscle cells. Some are made to be bone cells, hair cells, or nail cells. The entire body is a collaboration of cells working in a union to function the body. These cells are not just a thing, they move. As they move they create a frequency. Anything that moves creates energy. Think of the running river that is harnessed as power for electricity, or the original windmill for the same purpose. Now, we see large turbines collecting the power of wind. It is not hard to see that we are a mighty source of energy. EVEN THOUGH it is not energy you can see! If the body is simply the outward expression of one's essence, then this produced energy and the energy of the essence of oneself must be in vibratory resonance or there is discord; physically, mentally, emotionally, spiritually; in whichever way will be most effective to make you stand up and take notice.

The essence will always fight to remain in vibratory resonance. Anything else causes illness, stress, and discomfort. If one feels the strains of these three, then you can be assured that the energies of the physical and essential are not in alignment with each other.

As is stated in the first law of physics, energy will never be killed or shut off; as a universal law, it must transform, morph, move and genesis. Also, a source of energy cannot be removed without its space being filled up with something new. The same pool of energy has been transforming, changing, recycling and regenerating since the beginning of time. We are all made up of this original energy, even though we are in our own separate forms.

When spirit being is incarnated into substance as a human, one brings these blind memories of the cosmic essential energy with them. These emotions and vibrations and patterns that incarnate, do so also into a physiological, scientific particle make up that holds it within itself (the form of the body). As these physical cells grow, transform, and change, they repeat cycles based on the memory of makeup that they store within themselves. This is why taking a drug to relieve oneself of a particular ailment works temporarily but it is only temporarily because it works on the present cells relieving them of the SYMPTOM, but not of the diseased pattern, so the newly formed cells are constructed off the same pattern of the adult cell. All of it is there and one together: the physical construct of cells which form the tissues, which from the organs, which from the systems, which form the body, and the metaphysical construct of energetic and vibratory patterns of the entire stock of thoughts, beliefs, concepts, and ideals throughout time. What has been experienced before will be experienced again; it can only be up to the individual to determine the experience it will choose to have, which is why it is imperative to understand that, "You have much more power than ever believed to influence your physical and mental realities. Your mindset is recognized by your body — right down to the genetic level, and the more you improve your mental habits, the more beneficial response you'll get from your body. You can't control what has happened in the past, which shaped the brain you have today, programmed your cells, and caused certain genes to switch on. However, you do have the power in this moment and going forward to choose your perspective and behavior, which will change your brain, cells, and genes."[54]

Vibration

So far, we can understand that changing the frequency and vibrational patterns that affect the cell on every level of existence, as far as an individual experiential process goes, can be accomplished through awareness. But what of the epigenetic and cosmic original frequencies?

More needs to be studied to understand how to, if it is possible, and it must be as energy is always growing, changing and adapting, it must be believed that it is possible to change the original frequency that came from the cosmic, but the code which comes from your physical cells and the imprints of your

[54] http://www.twitter.com/dlhampton

life experiences is and can be regularly manipulated. This determination of destiny begins to construct the legacy you leave behind. Changing the original frequency, the cosmic, the essential is only changed by a beingness which raises the vibration so that only these higher vibrations are now a match for it. Higher vibrations attract higher vibrations, gradually recycling the lower vibrations making them more elevated and evolved. This is how old lower vibratory patterns are healed, and possibly could be all that is needed for changing ANY frequency, even those of the cosmos and epigenetics.

"Vibrational energy healing, or harmonic healing, dates back to the ancient civilizations of the Lemurians, Aztecs, Egyptians and the Chinese. According to Gurudas, author of Gem Elixirs and Vibrational Healing, vibrational remedies are 'tinctures of liquid consciousness' containing an evolutionary force in the shape of a particular energy pattern."[55]

"The field of the individual receiving these healing

[55] "Vibrational Energy Healing." Holistic MindBody Healing, www.holistic-mindbody-healing.com/vibrational-energy-healing.html.

vibrations entrains and resonates with that of the healer. Negative states may shift immediately or the process may take time over multiple sessions. It usually takes a while for the field to stabilize and hold a higher frequency.

After a session, it is not uncommon for the field to go back to its previous frequency,"[56]   Is the currently held belief on vibrational healing. While true, in essence, we also must be wary of suggestions that we are allowing access to the individual to accept. We must speak only of that which we wish to transform, and understand that healing cannot happen without the participation of all beings involved in the healing. It is not a task that can be done outside of someone, but in fact is guiding the energy which is brought by the individual. Help can be made to create a higher vibrational environment, but ultimately the healed is the healer, themselves. They need to have an active participation in the course of the conduction to make a lasting impact. It is impossible for ANY healer to

---

[56] "Vibrational Energy Healing." Holistic MindBody Healing, www.holistic-mindbody-healing.com/vibrational-energy-healing.html.

have more knowledge than the wisdom of the own cells of the body that is in need of healing.   Yes, a healer can lay the wires, but the electricity won't travel through it unless connections are made for travel.

Healing

The Place for Positive Thinking

Understanding the multiple avenues through which we can receive our genesis coding imprints, we can side with the scientists and doctors in belief that it is not just "positive thinking" that causes healing to happen. There have been miracle stories where this seems to be the case, and there have been stories where positive thinking was present but the disease progressed anyway. The theories of the Genesis Code and the Conceptual Cell, when understood in the context that the coding is sourced from a multitude of causal energies, can begin to define why positive thinking, while definitely having an impact in the right direction, does not fully explain healing. This is partly what has brought about the great divide between medical doctors and alternative practitioners: the fact that there are many documented cases where medicine prevails, but also many documented cases where the same situation does not result in health. The same is true in the case studies of alternative practitioners. This is suggestive of the tenement of the Conceptual Cell Theory that science and theology cannot be separated, as well stated by Dr. Samuel Hahnemann, "You cannot divorce medicine and theology. Man exists all the way down from his innermost spiritual to his outermost natural." In "The Lesser Writings of Dr. Samuel Hahnemann".

In truth, theories involving positive thinking are effective, though they are only partially correct.

True and permanent healing involves more than positive thought, which is only a piece of it, and only in use as a magnet. If these positive vibrational patterns were not already coded into the cells, there would not be these attractive energy protons to attract from the outer world.

Much is talked about as far as rewiring the brain for positive thinking, but is the mind inherently already wired for positive thinking? It could be said that could be determined in part by their epigenetic disposition. Years and years ago, it would seem that the mind was originally inherently wired for positive thinking, at least according to Dr. Samuel Hahnemann who said, "As long as man continued to think that which was true and held that which was good to the neighbor, that which was upright and justice so long man remained free from disease, because that was the state in which he was created."[57]

He was also known to believe that, "The body became corrupt because man's interior will became corrupt."[58]

The fact of the matter is that whether the brain is inherently wired for positivity or not, it CAN be RE-wired. There have been many theorists that say the mind's job is to look for problems. As a mechanism based on thought, it always needs something to solve. The human life offers up much for experience. However, it is how we perceive and adapt that makes a difference in our genesis coding. According to William Chen, "The reason is quite simple: They

[57] "The Lesser Writings" by Samuel Hahnemann
[58] The Lesser Writings" by Samuel Hahnemann

(our minds) are actually wired to pay more attention to negative experiences. It's a self-protective characteristic. We are scanning for threats from when we used to be hunter and gatherers."[59]

Fortunately we can rewire this patterning. Nerve cells, or neurons in your brain communicate through synapses. Every time one activates nerve synapses, the connections become more durable, reliable, and faster. Basically, the more you practice something, the easier it becomes without even thinking about it. Because of the brain's plasticity, and through practice, these connections are the reason the brain can be trained to do anything new. The conscious mind stores like a desktop on a computer all of the files it currently uses and continues to build on, so that they are at the ready to be brought out and used again as needed. As ideas, thoughts, or mind patterns are no longer used as often, they are pushed

[59] Chen, Written by Walter, and Brian Bailey Read more. "How To Rewire Your Brain for Positivity and Happiness - The Buffer Blog." Social, 11 Mar. 2016, blog.bufferapp.com/how-to-rewire-your-brains-for-positivity-and-happiness.

down and stored in the internal memory, the subconscious mind. It never deletes or removes any of these files. So all memories in your past history are able to be called up at any time. Because the subconscious mind is also the truth mind, and the essence energy, all ideas, thoughts, concepts, possibilities of the entire history of the entire essence, you can call upon any answer you need.

Warren Chen goes on to explain that, "We can harness the brain's plasticity by training our brain to make positive patterns more automatic. When we practice looking for and being more aware of positive aspects of life, we fight off the brain's natural tendency to scan for and spot the negatives. Naturally we bring ourselves into better balance." [60]

Much research has been conducted on the brain and intelligence. But even the brain needs to be looked at beyond the physical structure. The MIND has also been a topic of

---

[60] Chen, Written by Walter, and Brian Bailey Read more. "How To Rewire Your Brain for Positivity and Happiness - The Buffer Blog." Social, 11 Mar. 2016, blog.bufferapp.com/how-to-rewire-your-brains-for-positivity-and-happiness.

theory and in the mind is where the power lies, at least according to those that believe in positive thinking. The interesting thing to note is that, again, the mind is only powerful according to our own perceptions and belief in it. Tenets of the Conceptual Cell Theory tell us that not only is our cellular patterning based in Genesis Coding(conception; individually and universally), but also in Conceptual beliefs, thoughts, and ideas. The mind is key but it is only powerful based on our own ideas, thoughts, and beliefs, and/or our ideas, thoughts, and beliefs over what we are believing and having ideas over. It is also prone to the power of the beliefs, thoughts, and ideas we are subjected to in our environments, and in our epigenetics, as already mentioned. It comes down to power; who has it, what has it, and where it is. Furthermore, often we allow it to BECOME truly powerful even though it did not originate that way. It only originated to be in power in one's own MIND. One believed it to be so. One believed oneself to be able to think themselves out or into a situation. Instead of using it as a tool for the essence, it is used as a power a god that rules YOU.

This powerful entity, is given complete authority and we are a culture that determines itself on sight, not vision, and proof. Yet, the mind cannot be proven. There are no scientific studies that can pick out the mind and study it under a microscope, try as they might. "As a general rule, scientists consider something true only when it can be meaningfully observed or measured. The unconscious mind, by definition, can't be. After all, its central feature is that it's completely inaccessible."[61]

The Conscious, Subconscious, and Unconscious Mind

From Freud's theories more than a century ago, the power of the mind, the unconscious, mind, and the subconscious mind has been evaluated, as well as it's

[61] Feldman, David B. "Does the Unconscious Really Exist?" Psychology Today, Sussex Publishers, 17 July 2017, www.psychologytoday.com/blog/supersurvivors/201707/does-the-unconscious-really-exist.

potential power over us and our actions. Much has been discussed about the mind having a "homunculus", "little man", or an Id Ego as Freud suggests that directs us what to do. Much has been researched on the automation of the mind's activities. But the real interesting aspect to note that according to David B. Feldman, a professor of counseling psychology at Santa Clara University, "A key difference between the classical unconscious and this more modern perspective is the degree to which automatic processes are accessible and therefore changeable. In Freud's view, not only was the unconscious impossible to directly observe, we were utterly at its mercy. Automatic thoughts and behaviors, on the other hand, are much easier to access. We can consciously tune into our daily commute simply by wanting to. We can even choose to take a different route. Likewise, my client was able to access her negative automatic thinking just by paying attention to it, even though initially she wasn't fully aware it was there. That's one of the reasons that psychologists are increasingly teaching mindfulness meditation — among many other advantages of such practices, they can help clients tune into thoughts they previously weren't noticing. Of course, changing our automatic thoughts isn't nearly as easy as tuning into them. Just as playing a sonata or riding a bike takes conscious repetitive practice to become automatic, changing our ingrained thinking patterns requires similar dedication and practice."[62]

What Professor Feldman speaks of supports one of the Genesis Code Theory's tenets that becoming aware of the activities of the mind can be powerful in altering the code the cells operate off of. But there is one fault of the mind. It is an entity OUTSIDE of and operating outside of the human soul, being the fact that "thinking" is not "knowing". So the question begs to be answered: does the mind operate out of itself? Or is it guided by something? Is this where "soul"

[62] Feldman, David B. "Does the Unconscious Really Exist?" Psychology Today, Sussex Publishers, 17 July 2017, www.psychologytoday.com/blog/supersurvivors/201707/does-the-unconscious-really-exist.

comes in? And if so, it proves the Genesis Code Theory's basic belief, that science cannot be separated from theology. Healing cannot happen in a hospital room void of contact with the soul, and all of the cosmic energies that make it up.

The Existence of a Soul

"Discovering the truth about the existence of the soul has been a topic long known to researchers and theologians."[63]

"It's said to be the ultimate animating principle by which we think and feel, but isn't dependent on the body. Many infer its existence without scientific analysis or reflection. Indeed, the mysteries of birth and death, the play of consciousness during dreams (or after a few martinis), and even the commonest mental operations – such as imagination and memory – suggest the existence of a vital life force – an élan vital – that exists independent of the body….Everything

---

[63] Vinita. "WHAT IS THE SOUL? EVIDENCE OF ITS EXISTENCE." Health Consciousness, 20 Oct. 2017, www.thehealthconsciousness.com/soul-evidence-existence/.

knowable about the "soul" can be learned by studying the functioning of the brain. In their view, neuroscience is the only branch of scientific study relevant to understanding the soul...the current scientific paradigm doesn't recognize this spiritual dimension of life. We're told we're just the activity of carbon and some proteins; we live awhile and die. And the universe? It too has no meaning. It has all been worked out in the equations – no need for a soul. But biocentrism – a new 'theory of everything' – challenges this traditional, materialistic model of reality."[64]

Biocentrism

Dr. Robert Lanza, biomedical researcher, is considered the "standard bearer for stem cell research" and gained the title when he has proven that stem cells can provide essentially a way back to inherently healthy cells producing healthy

---

[64] Lanza, Robert. "Does The Soul Exist? Evidence Says 'Yes'." Psychology Today, Sussex Publishers, 21 Dec. 2011, www.psychologytoday.com/blog/biocentrism/201112/does-the-soul-exist-evidence-says-yes.

manifestations.[65] Lanza introduces a radical new thought with his theory of biocentrism. "Biocentrism completes this shift in worldview, turning the planet upside down again with the revolutionary view that life creates the universe instead of the other way around. In this new paradigm, life is not just an accidental byproduct of the laws of physics. Lanza proposes a biocentrist theory which ascribes the answer to the observer rather than the observed. The work is a scholarly consideration of science and philosophy that brings biology into the central role in unifying the whole."

Whether one morally believes in the use of embryonic human stem cells, the facts prove that a "master cell" exists. That there IS something as the Conceptual Cell Theory suggests, that the blueprint and coding of the cells can be manipulated, at least biologically. "Human embryonic stem cells are biological shape shifters. They uniquely possess the ability to become any other cell in the body, which makes

---

[65] "The great stem cell dilemma." Fortune, fortune.com/2012/09/28/the-great-stem-cell-dilemma/.

them extraordinarily powerful and dangerous. When injected in an undifferentiated state or left unchecked in a petri dish, they form disgusting teratoma tumors consisting of many features — hair, skin, teeth, etc. But when properly directed to a desired outcome, they become healthy liver cells or heart muscle, wrinkle-free facial skin, or you name it. "The power of an embryonic stem cell is it's the master cell," says Lanza."[66]

The idea that there is a master cell, at all suggests that there is indeed a "blank slate" reliant on the programming, from wherever it comes. With careful guidance it becomes a beautiful healing modality. With improper and poor guidance, it becomes a thing of disease and disgust. Much like we see in the world today: it's in the programming, which seems to prove the "nurture" side of the theory, though because it relates to biology and cellular structure, it also proves the "nature" side.

---

[66] "Robert Lanza » Biocentrism / Robert Lanza's Theory of Everything." Robert Lanza RSS, www.robertlanza.com/biocentrism-how-life-and-consciousness-are-the-keys-to-understanding-the-true-nature-of-the-universe/.

Function of the Mind

As powerful as it has been allowed to be, the mind has faults of efficacy, of intelligence, of decisiveness, because it doesn't know KNOWING it knows THINKING. The observer converts thinking into knowing. Consider the fact that, "Experiments make it increasingly clear that even mere knowledge in the experimenter's mind is sufficient to convert possibility to reality. Consider the famous two-slit experiment. When you watch a particle go through the holes, it behaves like a bullet, passing through one slit or the other. But if no one observes the particle, it exhibits the behavior of a wave and can pass through both slits at the same time. This and other experiments tell us that unobserved particles exist only as 'waves of probability' as the great Nobel laureate Max Born demonstrated in 1926. They're statistical predictions – nothing but a likely outcome. Until observed, they have no real existence; only when the mind sets the scaffolding in place, can they be thought of as having duration or a position in space."[67] This proves that one's reality is based in the mind and the belief in possibility. For disease to exist, the wrong pattern of possibility has been initiated.

Further complicating the idea is that thinking cannot perceive ALL POSSIBILITY. It can only perceive to the capacity of the man. And because man is not GOD, or the entirety of the essence itself, it's scope of possibility is limited. On top of that, each set of possibilities that can be thought of by one man does not encompass the knowledge of the whole.

Individual Human Experiences in the Context of the Whole

As the humans use physical matter to create and recreate cellular structures again and again, the energy patterns come along with that. The energy patterns, remember that

---

[67] Lanza, Robert. "Does The Soul Exist? Evidence Says 'Yes'." Psychology Today, Sussex Publishers, 21 Dec. 2011, www.psychologytoday.com/blog/biocentrism/201112/does-the-soul-exist-evidence-says-yes.

have been pooled and transformed for centuries since the beginning of time. The energy patterns that have a memory and imprint off of each other. Energy is all the same through in and through out. What one part of energy experiences, the whole experiences as well. So this part of one that is energy, this part that originated from this same pool comes with it all of the memories, the thoughts, the beliefs, the emotions, the knowledge that has been gained through all of its successive incarnations in every form possible.

The code that is inside of one being holds all memories of the whole within the cosmic essential energy, as well as the imprints of the physical cells that make up the body, and all of the memories, emotions and experiences that are imprinted within the essence of the human experience. The "human condition" is in fact better named the "human experience" that makes us who we are.

Conclusion

The totality of healing is the purpose and function of the Genesis Code. The truth of the legacy is that our efforts span much farther the reach of one individual, but wisdom and higher vibrational energies for generations to come.

All modalities have been developed, because someone saw the benefit of something that matched that particular level of the person at that space in time. All methods and wisdom should be used, but never to the exclusion of the code of the Genesis, which is the construct, the map of the entire imprint.

Gaining an understanding of true healing brings us to what one is truly meant for, our own function and purpose, where true healing exists; the return to Satori, to Oneness, to wholeness, in which the cells can develop their blueprint within a framework of high vibrational energy. Not only for themselves, but in the frequency emanating from them. True healing cannot simply be for one individual, because as mentioned before, we are not individuals, separate and without care. We are part of the fabric of the entirety . If we are to look as Dr. Hahnemann proved, at his Totality of Symptoms because one symptom cannot simply make up the whole of a person, the same is true of our cosmic representation and inheritance.

Science cannot be made individual and medicine cannot erase the truth of energy if true healing is ever to be understood. In every scope, each individual through their actions are determining the course of the next generations and the direction the next centuries will take according to the path of humanity. As the vibration elevates higher and higher, a world of lower actions, mentalities, behaviors, and beliefs simply will not exist. The effort and intention, then, falls to elevating each individual's plane of existence and vibrational contribution to the whole. This is done through careful, compassionate examination of the conceptual cell with the idea of the Genesis code in mind by improving health thru individual and community platforms, and a shift in procedures that involve pulling what's within - out, rather than current procedure of attack from outside killing the in.

The evolution of humanity will not be healthfully moved forward into the next generation without understanding this. Science cannot divorce Energy from healing. By law it simply cannot.

References

Team, The MNT Editorial. "What Is Medicine? A History Of Medicine." *Medical News Today*, MediLexicon International, 5 Jan. 2016, www.medicalnewstoday.com/info/medicine.

C., German. "Kinetic Energy and Potential Energy." *Kinetic and Potential Energy | Wyzant Resources*, www.wyzant.com/resources/lessons/science/physics/kinetic-and-potential-energy.

Understanding Evolution Team. "Evolution at Different Scales: Micro to Macro." *Evolution at different scales: micro to macro*, Understanding Evolution, evolution.berkeley.edu/evolibrary/article/evoscales_01.

Spees JL, Olson SD< Whitey MJ, Prockop DJ. Mitochondrial transfer between cells can rescue aerobic respiration. PNAS. 2006; 103(5): 1283-1288.

https://www.coursehero.com/file/19960291/nature-vs-nurture/

Team, The MNT Editorial. "What Is Modern Medicine?" *Medical News Today*, MediLexicon International, 5 Jan. 2016, www.medicalnewstoday.com/info/medicine/modern-medicine.php.

"What is Quantum Energy Transformation™ (QET™)?" *What is Quantum Energy Transformation™ (QET™)? |*, www.quantumhealingcenter.com/quantum-energy-transformation.

Dutton, P L, and C C Mosser. "Quantum biomechanics of long-Range electron transfer in protein: hydrogen bonds and reorganization energies." *Proceedings of the National Academy of Sciences of the United States of America*, U.S. National Library of Medicine, 25 Oct. 1994, www.ncbi.nlm.nih.gov/pmc/articles/PMC44996/.

www.zen-buddhism.net. "Satori or Awakening." *ZEN BUDDHISM | Zen Concepts | Satori*, www.zen-buddhism.net/zen-concepts/satori.html.

https://en.wikipedia.org/wiki/Germ_theory_of_disease

"Intro." *Dr. Thurman Fleet*, www.drthurmanfleet.com/.

"About Us." *CONCEPT-THERAPY*, concept-therapy.org/portfolio-item/about-us/.

http://www.wholehealthnow.comhomeopathypro/miasms-03.html

Phillips, Theresa. "How Genetic Polymorphism Promotes Diversity and Lasts Over Generations." The Balance, www.thebalance.com/genetic-polymorphism-what-is-it-375594.

Phillips, Theresa. "How Genetic Polymorphism Promotes Diversity and Lasts Over Generations." The Balance, www.thebalance.com/genetic-polymorphism-what-is-it-375594.

"How do the laws of thermodynamics provide evidence for creationism?" CompellingTruth.org, www.compellingtruth.org/laws-of-thermodynamics.html.

"What is biogenesis?" CompellingTruth.org, www.compellingtruth.org/what-is-biogenesis.html.

"What is the meaning of creation 'ex nihilo'?" CompellingTruth.org, www.compellingtruth.org/creation-ex-nihilo.html.

Hampton, Debbie. "How Your Thoughts Change Your Brain, Cells and Genes." The Huffington Post, TheHuffingtonPost.com, 23 Mar. 2016, www.huffingtonpost.com/debbie-hampton/how-your-thoughts-change-your-brain-cells-and-genes_b_9516176.html.

http://www.twitter.com/dlhampton

"What Is Epigenetics?" Edited by Steven Dowshen, KidsHealth, The Nemours Foundation, Jan. 2014, kidshealth.org/en/parents/about-epigenetics.html.

"Hering's Law of Cure." Healing Naturally by Bee, www.healingnaturallybybee.com/herings-law-of-cure/.

Chen, Written by Walter, and Brian Bailey Read more. "How To Rewire Your Brain for Positivity and Happiness - The Buffer Blog." Social, 11 Mar. 2016, blog.bufferapp.com/how-to-rewire-your-brains-for-positivity-and-happiness.

Feldman, David B. "Does the Unconscious Really Exist?" Psychology Today, Sussex Publishers, 17 July 2017, www.psychologytoday.com/blog/supersurvivors/201707/does-the-unconscious-really-exist.

Lanza, Robert. "Does The Soul Exist? Evidence Says 'Yes'." Psychology Today, Sussex Publishers, 21 Dec. 2011, www.psychologytoday.com/blog/biocentrism/201112/does-the-soul-exist-evidence-says-yes.

"The great stem cell dilemma." Fortune, fortune.com/2012/09/28/the-great-stem-cell-dilemma/.

"Robert Lanza » Biocentrism / Robert Lanza's Theory of Everything." Robert Lanza RSS, www.robertlanza.com/biocentrism-how-life-and-consciousness-are-the-keys-to-understanding-the-true-nature-of-the-universe/.

"Vibrational Energy Healing." Holistic MindBody Healing, www.holistic-mindbody-healing.com/vibrational-energy-healing.html.

"Mitochondrial DNA - Genetics Home Reference." *U.S. National Library of Medicine*, National Institutes of Health, ghr.nlm.nih.gov/mitochondrial-dna.

"BibleGateway." *Philippians 4:6-9 KJ21 - - Bible Gateway*, www.biblegateway.com/passage/?search=Philippians%2B4%3A6-9&version=KJ21.

"Inherited Tendencies." Center for Homeopathy, www.centerforhomeopathy.com/blog/inherited-tendencies.

"How Your Thoughts Program Your Cells." High Existence, 7 Feb. 2011, highexistence.com/thoughts-program-cells/.

Sassandahalf, et al. "How Your Thoughts Change Your Brain, Cells, And Genes." The Best Brain Possible, 1 Dec. 2017.

"Hering's Law of Cure." Healing Naturally by Bee, www.healingnaturallybybee.com/herings-law-of-cure/.

Quanta and Wave-Particle Duality - Quantum Theory and the Uncertainty Principle - The Physics of the Universe, www.physicsoftheuniverse.com/topics_quantum_quanta.html.

"Vibrational Energy Healing." Holistic MindBody Healing, www.holistic-mindbody-healing.com/vibrational-energy-healing.html.

Resources

Team, The MNT Editorial. "What Is Modern Medicine? A History of Medicine." *Medical News Today*, MediLexicon International, 5 Jan. 2016, www.medicalnewstoday.com/info/medicine/modern-medicine.php.

http://physics.bu.edu/~duffy/py105/EnergyConservation.html

Wilson, Robert Anton. "Quantum Psychology: How Brain Software Programs You and Your World, Hilaritus Press, 23 Aug. 2016.

Sailus, Christopher. "What is Biomechanics? - Definition & Applications." *Study.com*, Study.com, study.com/academy/lesson/what-is-biomechanics-definition-applications.html.

http://www.drthurmanfleet.com/teachings/

https://www.thebalance.com/genetic-polymorphism-what-is-it-375594
Plevell, Amanda. "The Genesis Code" Flight Plan Publishing, 2013.

http://kidshealth.org/en/parents/about-epigenetics.html

Laestadius, Lars L. The lunatic: an insight into the order of grace: commonly known as "Servant in the crazy house": systematically presented in the form of observations of the characteristics and states of the soul, in accordance with the psychological prespectives of the biblical authors, pertaining to the highest idea of christianity - reconciliation. Biblioteca Laestadiana, 2015.

http://www.centerforhomeopathy.com/blog/inherited-tendencies

www.thebestbrainpossible.com/how-your-thoughts-change-your-brain-cells-and-genes/.

"Hering's Law of Cure." Healing Naturally by Bee, www.healingnaturallybybee.com/herings-law-of-cure/.

Huang, Kerson, and Rosemary Huang. I ching. Workman Pub., 1987.

Quora. "What Is the Science Behind Hypnosis?" The Huffington Post, TheHuffingtonPost.com, 3 June 2015, www.huffingtonpost.com/quora/what-is-the-science-behin_b_7505012.html.

"The reality of truth" YouTube Documentary, reinforces this at about the 3:22 marker

About the Author

Amanda E. Soulvay Plevell was born Amanda Elaine Dahl in Montevideo, MN, USA in 1977. After graduating high school, she pursued and obtained a Bachelor's degree in Education and Applied Psychology: Counseling at St. Cloud State University in St. Cloud, MN. Post graduation Ms. Plevell continued education from the time of her graduation in 1998 to the present. She earned a Certification as a Natural Health Practitioner in 2007, along with a certification as a trained EDS Technology Technician. Continuing on she earned a Diploma in Hair, Tissue, Skin, Mineral Analysis from the College of Natural Medicine, Larnaca, Cyprus in 2017 and a Diploma in Biochemical Nutrition also from the same institution. She most recently completed a Doctorate in Clinical Natural Medicine/Naturopathy and PhD in Natural Medicine, both from New Eden School of Clinical Natural Health in Indiana in 2017.

Professionally, Amanda taught as a teacher before working at the St. Cloud Hospital's Inpatient Pharmacy. She is the owner and Founder of the Natural Source for Integrated Wellness, a teaching clinic for Consciousness Lifestyle, Community Wellness with intent on expanding global good through natural functional wellness, lifestyles by intentional design,and concept pathology The Natural Source heads up system specific online wellness support programs and business development for the Next Generation of Healers.

She has currently published nearly thirty books on the topics of wellness, greatness, and success conditioning. She is a trainer, speaker, community wellness and teen advocate developing programs to support teen growth and to encourage each individual she encounters to be the change.